PREFACE

The idea for this book came to me last year when my nephew and niece asked me to help them with their home work in construction, which they both had difficulty with. I assumed that they knew nothing about the topic and used my personal work, notes and questions to teach them. These materials were eventually used to make this Mathematics textbook a reality. After two days, they fully understood all the construction topics in the CSEC Mathematics syllabus and were able to do questions from past papers.

I joined the Teaching Service in Trinidad and Tobago in 1995 as a Teacher of Mathematics and in 1997 I got a part time job teaching Mathematics to repeater students at a private Secondary School. At one point in time I was working part time at three private Secondary Schools while I was full time at a Government Secondary School. With repeater students, a lot of work needed to be done in a short space of time, from September to April the following year in order to ensure that they improved their grades. About three to five years of work had to be covered in this short space of time, so I designed my lesson plans and methods of teaching in such a way that I assumed they knew nothing in Mathematics. This meant that a tremendous amount of work had to be done on the board and adequate home work given regularly and corrected in order to make sure that students learnt something. I started from basic concepts in all the CSEC Mathematics topics to make sure that most of the students understood the foundation work before proceeding to harder topics.

My lesson plans, assignments and methods of delivery helped me to be effective in the classroom and I was able to pitch the work for the mathematically weak students. Most repeater students often tell me that it was the first time that they actually understood certain topics. I have worked at public and private Secondary Schools in north, central and south Trinidad, so I got the opportunity to interact with students of all types of socioeconomic backgrounds with different intelligent levels.

At the private Secondary Schools where I worked at in the past, I was able to teach a wide array of subjects which broadened my view, knowledge and experiences. I taught Additional Mathematics, Biology, Human and Social Biology, Physics, Integrated Science, Geography, Commercial Numeracy, SAT Mathematics, GMAT Mathematics, GRE Mathematics, Introductory Statistics for the Behavioural Sciences, Introduction to Statistics for Psychology, Introduction to Quantative Methods for the ABE program, Computer Literacy, PowerPoint Presentation, Microsoft Word and Graphics Design. In China, I taught Oral English, Conversational English, Business English and Tourism English. I also have the knowledge to teach Social Studies, History, Principles of Business, Economics, Sociology, Psychology, International Relations and Politics.

This book will help form four, form five and repeater students to fully understand the construction topics covered in CSEC Mathematics. I purposely did not go beyond the scope of the CSEC Mathematics syllabus, because I do not want to turn off the students who are not doing Additional Mathematics. **The main feature of this book is that it was designed for students who want to quickly master construction at CSEC level.** A student who diligently works through the entire book and gets most questions correct will be able to do past paper questions without any difficulty. Students are advised to do over the incorrect problems to get the correct answer. In this way a student will know where the error is and how to avoid it next time.

I would like to thank my nephew and niece for giving me the idea for this book and other books in the Mathematics series. As a result, I wish to dedicate this book and the others in the series to my **nephews and nieces Christon, Joel, Natalia, Faith and Paris (Popo).**

CONTENTS

1 Vocabulary in construction

Arc   a curved line that is part of a circle.

Bisect   to cut into two equal parts. For example, to bisect a line, to bisect an angle, etc.

Bisector   the line that cuts an angle or another line.

Construct   this term automatically means to use a pair of compasses, a pencil and a ruler to show construction lines and arcs.

Draw   for example, "draw an angle of 100°" means that a protractor, a pencil and a ruler can be used instead of a pair of compasses.

Irregular polygon   a polygon in which the sides have different lengths and the angles have different sizes.

Line segment   a part or portion of a line with two end points.

A   B   C   D

The arrows represent a line. The line segments are AB, AC, AD, BC, BD and CD.
Properties of line segment:
1) A line segment is a path between two points.
2) It is named using its two end points.
3) A line segment has a definite length.

Perpendicular "at right angles"   a line that meets another at a right angle or at 90° is said to be perpendicular to it.

Perpendicular bisector   a line that passes through the mid point of a line segment and is perpendicular to it.

Polygon   this is a closed plane figure made up of three or more sides.

Quadrilateral   a four sided figure.

Regular polygon   a polygon that has all sides equal and all interior angles equal.

Semicircular arc   an arc that is half of a circle in length.

Sketch   a drawing that is used to represent a figure that is to be constructed.

2   Line segments

Use your ruler to measure the following line segments. Write the value next to each.

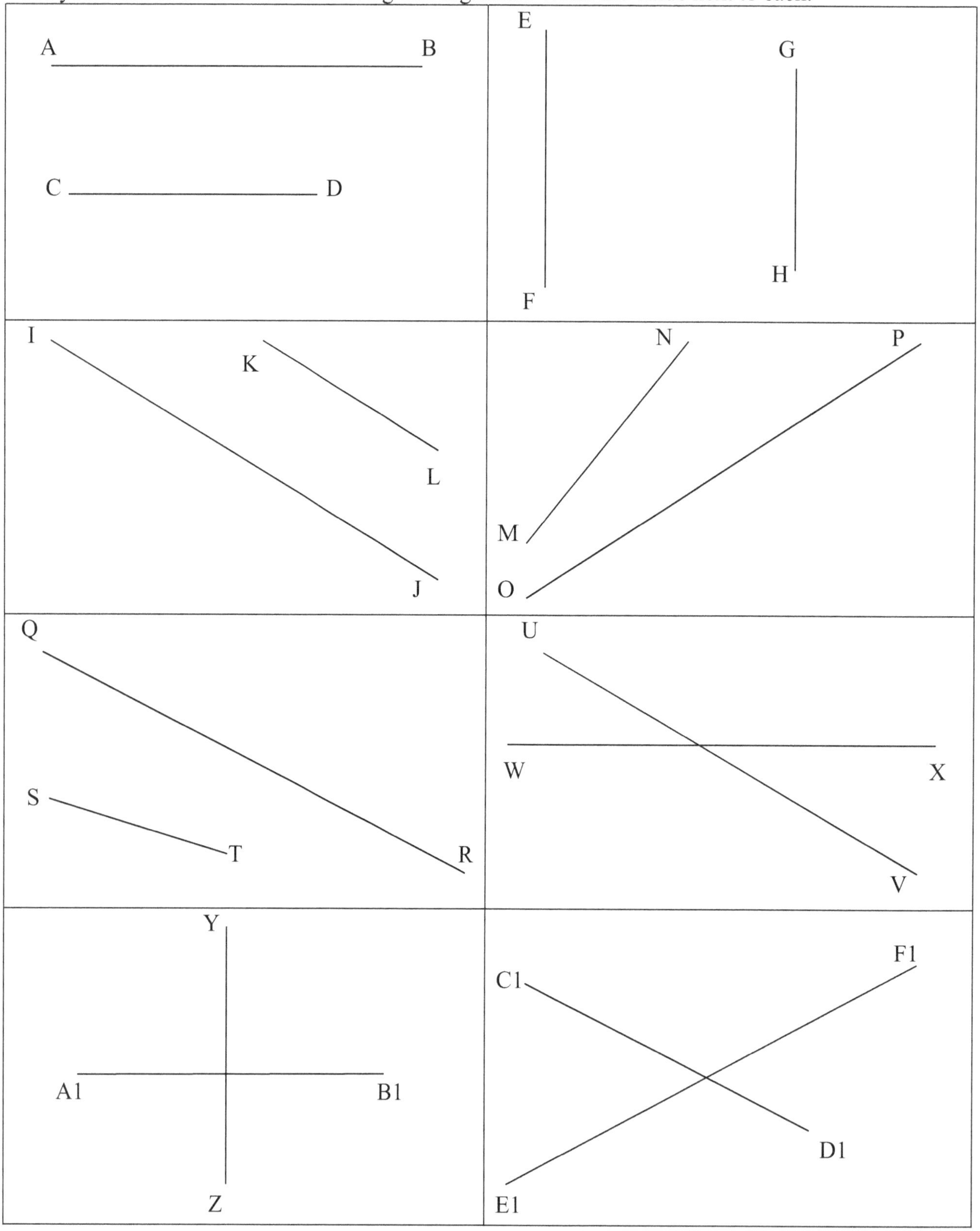

A _____ B

C _____ D

E

G

F

H

I

K

L

J

N

P

M

O

Q

S

T

R

U

W

X

V

Y

A1

B1

Z

C1

F1

D1

E1

Use your ruler and a pencil to draw and label the following line segments.

AB = 7 cm CD = 8 cm	EF = 4.5 cm GH = 7.5 cm
IJ = 3.7 cm KL = 4.8 cm	MN = 9 cm OP = 4.1 cm
QR = 60 mm ST = 45 mm	UV = 38 mm WX = 52 mm
YZ = 6.5 cm A1B1 = 3.8 cm	C1D1 = 22 mm E1F1 = 2.2 cm

Use a pair of compasses and a pencil to cut the lengths on the line segments.

E.g. Cut a value of 4 cm on the line segment.
STEPS:
1. Open the compass point and pencil point to 4 cm on your ruler.
2. Place the compass point on A and draw an arc on the line

A ——————————————— B

5 cm

C ———————————————————— D

6.5 cm
E
F

7 cm
H
G

25 mm
I
J

4.5 cm
K
L

7.5 cm
N
M

3.5 cm
O
P

3   To bisect a line segment

Bisect the following line segments using a pair of compasses and a pencil.

E.g.

Arc 2

Arc 1

Perpendicular
bisector

90°

A ———|———N———|——— B

AN = NB

C ————————————— D

E ————————— F

G————————————————H

I ————— J

K ————— L

Bisect the following line segments using a pair of compasses and a pencil.

A

B

C

D

E

F

G

H

I

J

K

L

Bisect the following line segments using a pair of compasses and a pencil.

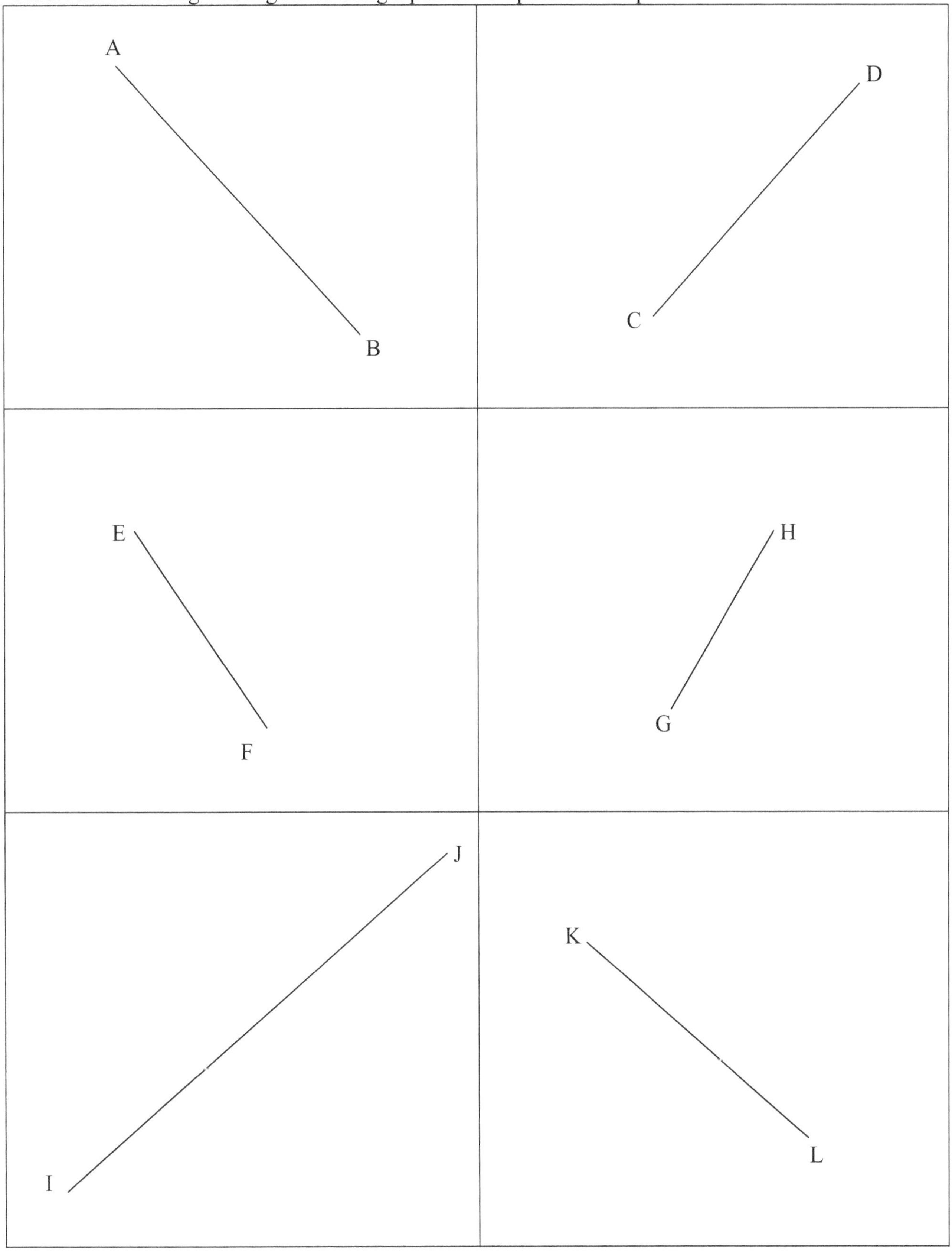

A

B

D

C

E

F

H

G

J

I

K

L

Draw the following line segments with your ruler and a pencil and bisect them using a pair of compasses and a pencil.

MN = 8 cm	OP = 6 cm
QR = 10 cm	ST = 7 cm

4   Parallel lines

Draw a line that is parallel to each in the following.

These lines are parallel, although they have different lengths. Parallel lines do not meet. They are the same distance apart.	

5 Perpendicular lines

Use a protractor to draw a line that is perpendicular to each in the following.

Perpendicular line

90°

A ———————— B   C ———————————— D

E   G

F   H

I   L

J   K

6   To construct a line perpendicular to another line from a point that is not on the line

1. Open your compass and pencil.
2. Place the compass point at A and the pencil
   point below the line LM.
3. Draw Arc 1.
4. From point C draw Arc 2.
5. From point D draw Arc 3.
6. Draw line 2.

Line 2

. A

Arc 1

L   C   D   M

Arc 2
Arc 3

. A

. A

. A

Construct a perpendicular to the given lines from point A.

Construct a perpendicular from point A to the base of the triangles.

E.g.

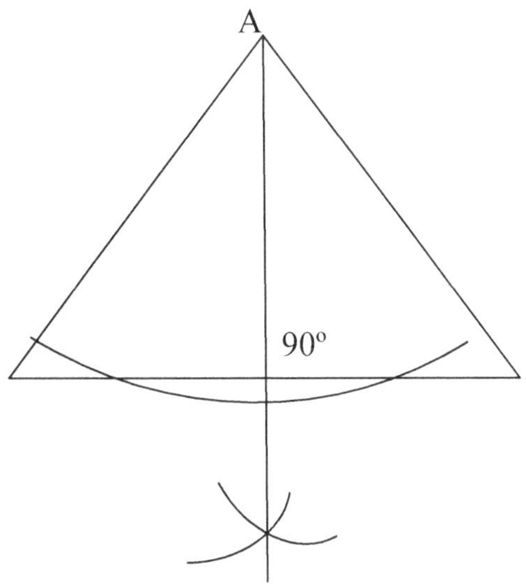

90º

Follow the instructions 1 to 6 in chapter 6.

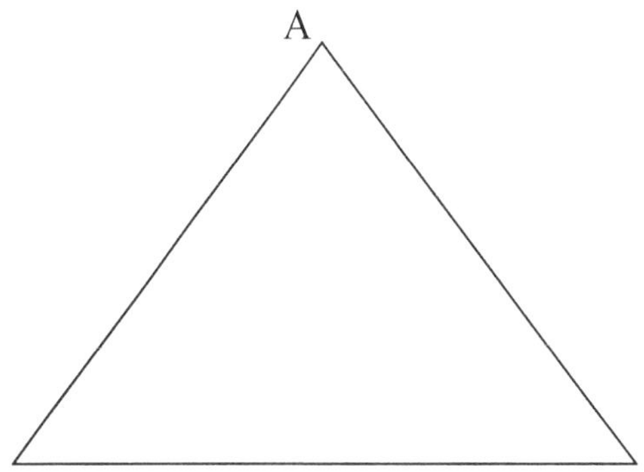

A

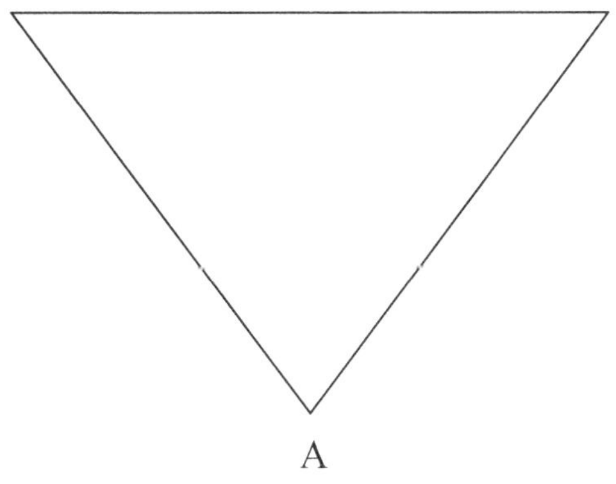

A

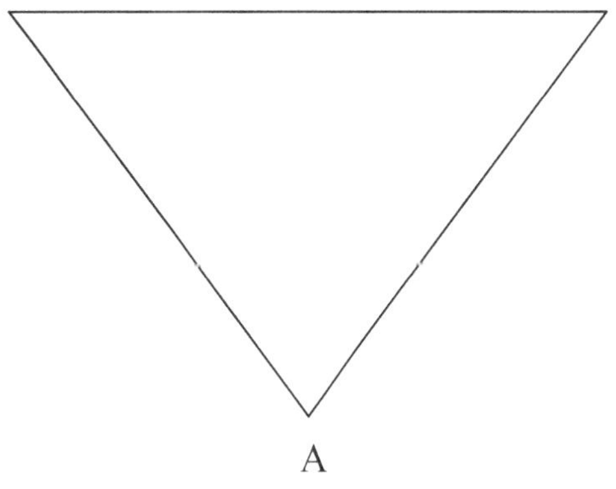

A

Construct a perpendicular from point A to the base of the triangles.

Construct a perpendicular from point A to the base of the triangles.

Construct a perpendicular from point A to the base.

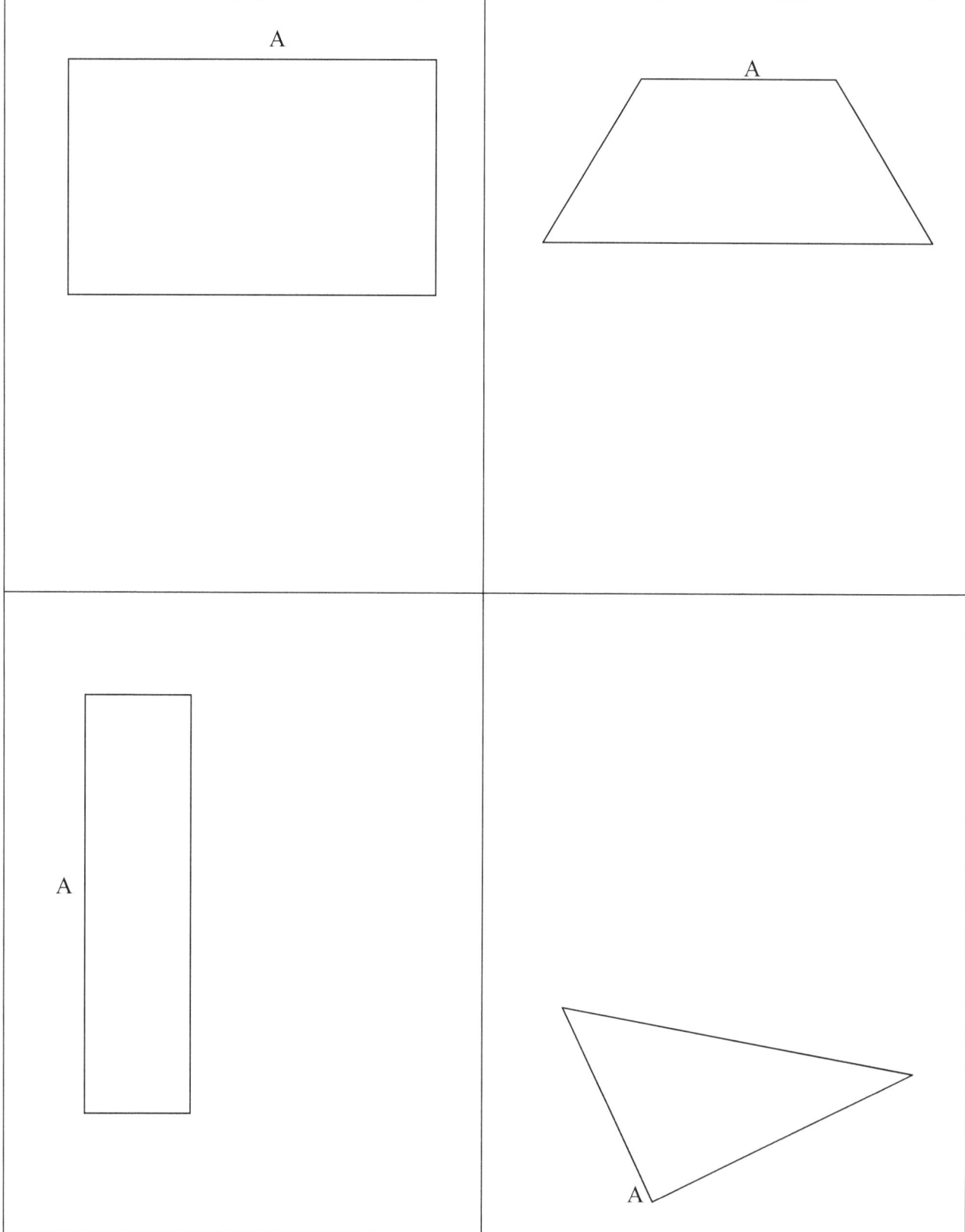

7   A protractor and its uses

A protractor is used to **measure** angles and to **draw** angles.

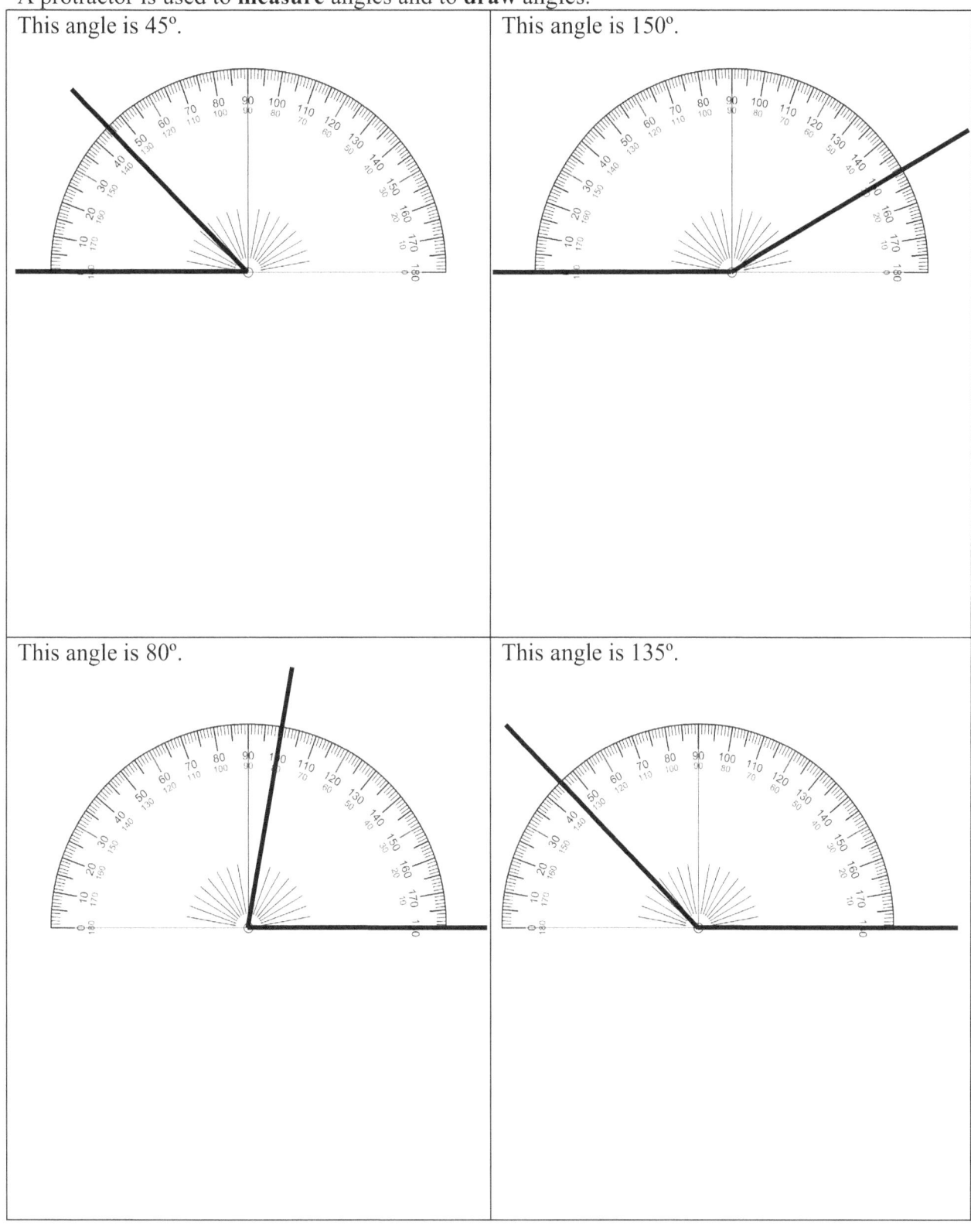

This angle is 45°.

This angle is 150°.

This angle is 80°.

This angle is 135°.

Measure the following angles and write their values.

23

Measure the following angles and write their values.

Measure the following angles and write their values.

Measure the following angles and write their values.

Measure the following angles and write their values.

Draw the following angles at the point A.

33°	80°
A ▬▬▬▬▬▬	A ▬▬▬▬▬▬
90°	60°
A ▬▬▬▬▬▬	A ▬▬▬▬▬▬
110°	135 °
A ▬▬▬▬▬▬	A ▬▬▬▬▬▬
145 °	170 °
A ▬▬▬▬▬▬	A ▬▬▬▬▬▬

Draw the following angles at the point B.

35°	70°
———— B	———— B
90°	65°
———— B	———— B
120°	145 °
———— B	———— B
165 °	170 °
———— B	———— B

Draw the following angles at the point C.

25° C ————	70° C ————
90° C ————	60° C ————
130° C ————	140 ° C ————
145 ° C ————	175 ° C ————

Draw the following angles at the point D.

25°	70°
_____ D	_____ D
90°	60°
_____ D	_____ D
135°	150 °
_____ D	_____ D
141 °	171 °
_____ D	_____ D

Draw the following angles at the point E.

20°	60°
E	E

80°	100°
E	E

145°	173°
E	E

Draw the following angles at the point F.

30°	70°
F	F
85°	105°
F	F
145°	163°
F	F

Draw the following angles at the point G.

32°	70°
G	G

81°	100°
G	G

145°	160°
G	G

Draw the following angles at the point H.

22°	50°
H	H
31°	105°
H	H
145°	165°
H	H

Draw the following angles at the point I.

22°	50°
31°	105°
145°	165°

8   How to bisect an angle

E.g. Bisect the given angle.
1. Open the compass and pencil to any length.
2. Place the compass point at A and draw Arc 1.
3. Place the compass point at B and draw Arc 2.
4. Place the compass point at C and draw Arc 3.
5. Draw a line from A through the point where Arc 2 and Arc 3 intersect.

E.g. Bisect the given angle.
1. Open the compass and pencil to any length.
2. Place the compass point at A and draw Arc 1.
3. Open the compass and pencil slightly. Your Teacher will tell you why. Place the compass point at B and draw Arc 2.
4. Place the compass point at C and draw Arc 3.
5. Draw a line from A through the point where Arc 2 and Arc 3 intersect.

Bisect the angle.

Bisect the angle.

Bisect the following angles.

Bisect the following angles.

Bisect the following angles.

Bisect the following angles.

Bisect the following angles.

Bisect the following angles.

Bisect the following angles.

Draw the following angles at A with your protractor and bisect them with a pair of compasses and a pencil.

30°	50°
A ———————	A ———————
80°	90°
A ———————	A ———————
120°	150°
A ———————	A ———————
168°	170°
——————— A	——————— A

Draw the following angles at B with your protractor and bisect them with a pair of compasses and a pencil.

30°	50°
——————————— B	——————————— B
80°	90°
——————————— B	——————————— B
120°	150°
——————————— B	——————————— B
168°	170°
——————————— B	——————————— B

46

Draw the following angles at C with your protractor and bisect them with a pair of compasses and a pencil.

20° C _____	40° C _____
85° C _____	90° C _____
130° C _____	140° C _____
150° C _____	160° C _____

Draw the following angles at D with your protractor and bisect them with a pair of compasses and a pencil.

30° ——————— D	40° ——————— D
50° ——————— D	80° ——————— D
120° ——————— D	130° ——————— D
150° ——————— D	160° ——————— D

Draw the following angles at E with your protractor and bisect them with a pair of compasses and a pencil.

36°	42°
E	E
54°	**84°**
E	E
128°	**160°**
E	E

Draw the following angles at F with your protractor and bisect them with a pair of compasses and a pencil.

36°	42°
F	F
54°	**84°**
F	F
128°	**160°**
F	F

Draw the following angles at G with your protractor and bisect them with a pair of compasses and a pencil.

34° G	40° G
58° G	88° G
150° G	162° G

Draw the following angles at H with your protractor and bisect them with a pair of compasses and a pencil.

24°	60°
H	H
38°	88°
H	H
140°	160°
H	H

9   Construction of 60°

E.g. Using a pair of compasses, a pencil and a
     ruler, construct a 60° at the point A.

1. Draw a line 1 of any length.
2. Open your compass and pencil to any length.
3. Place the compass point at A and draw Arc 1
   to cut at B.
4. Do not open or close your compass. The same
   length is needed for the next step.
5. Place the compass point at B and draw Arc 2.
6. Draw a line from A through the point where
   Arc 2 cuts Arc 1.
7. Measure the angle with a protractor and make
   sure that you got 60°.

Construct 60° at point A in the following.

A

A

A

A

A

Using a pair of compasses and a pencil, construct a 60° at the points A in the following.

10   Construction of 30°

E.g. Using a pair of compasses, a pencil and a ruler, construct a 30° at the point A.

STEP 1: Construct 60°. STEP 2: Bisect it.

1. Draw a line 1 of any length.
2. Open your compass and pencil to any length.
3. Place the compass point at A and draw Arc 1 to cut at B.
4. Do not open or close your compass. The same length is needed for the next step.
5. Place the compass point at B and draw Arc 2.
6. Draw a line from A through the point where Arc 2 cuts Arc 1 at C.
7. Place the compass point at B and draw Arc 3.
8. Place the compass point at C and draw Arc 4.
9. Draw a line from A through the point where Arc 3 and Arc 4 intersect. Measure to get 30°.

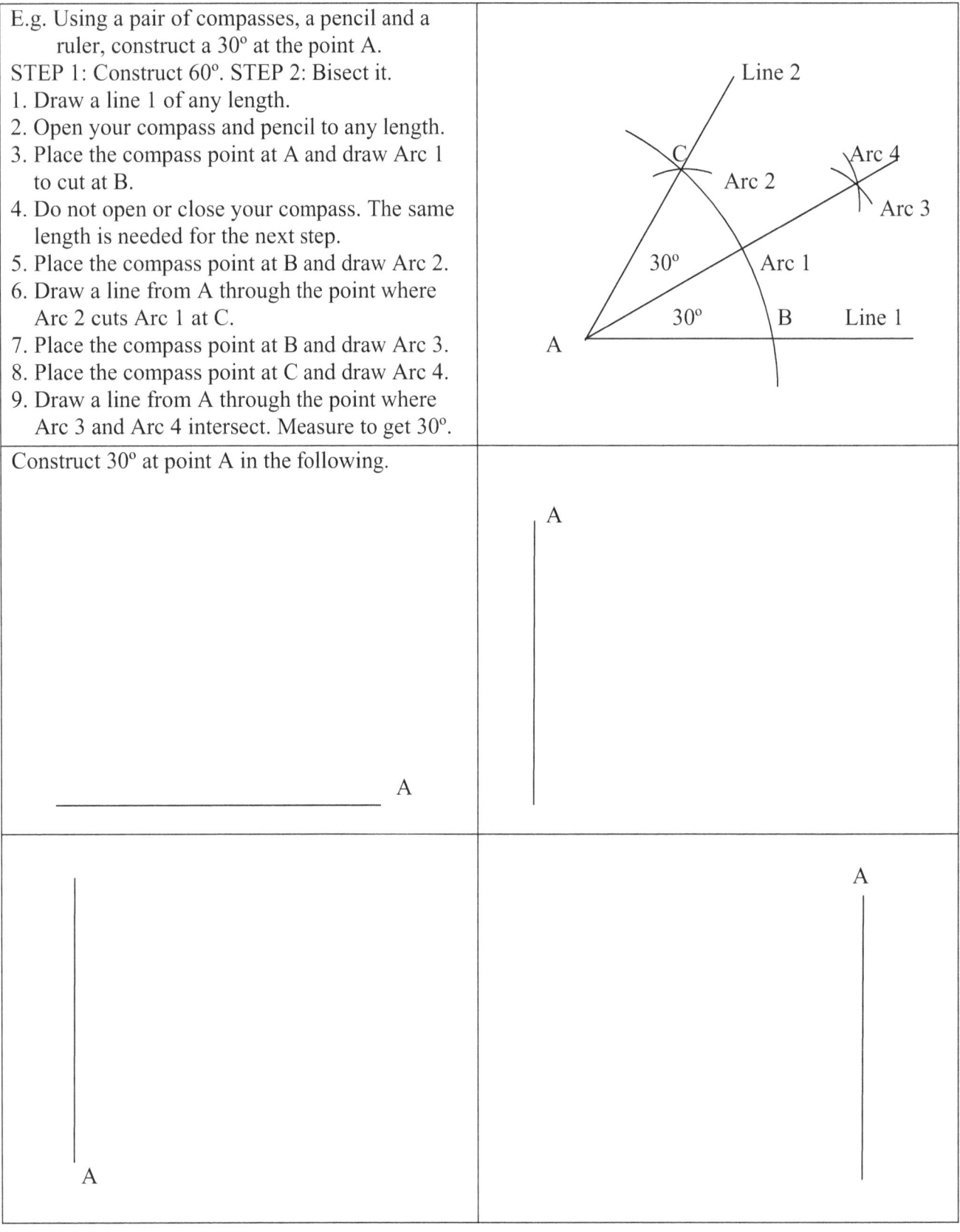

Construct 30° at point A in the following.

Using a pair of compasses, a pencil and a ruler, construct a 30º at the point A in the following.

REMEMBER to construct 60º first. THEN bisect it. A	A ———————————
——————————— A	A ——————————— A
A ——————————— A	A
A	A

11   Construction of 120°

E.g. Using a pair of compasses, a pencil and a
       ruler, construct 120° at the point A.

1. Draw a line 1 of any length.
2. Open your compass and pencil to any length.
3. Place the compass point at A and draw Arc 1
   to cut at B.
4. Do not open or close the compass.
5. Place the compass point at B and draw Arc 2.
6. Where Arc 2 cuts Arc 1, place the compass
   point there and draw Arc 3.
7. Place the ruler where Arc 3 cuts Arc 1 and
   draw a line to meet A.

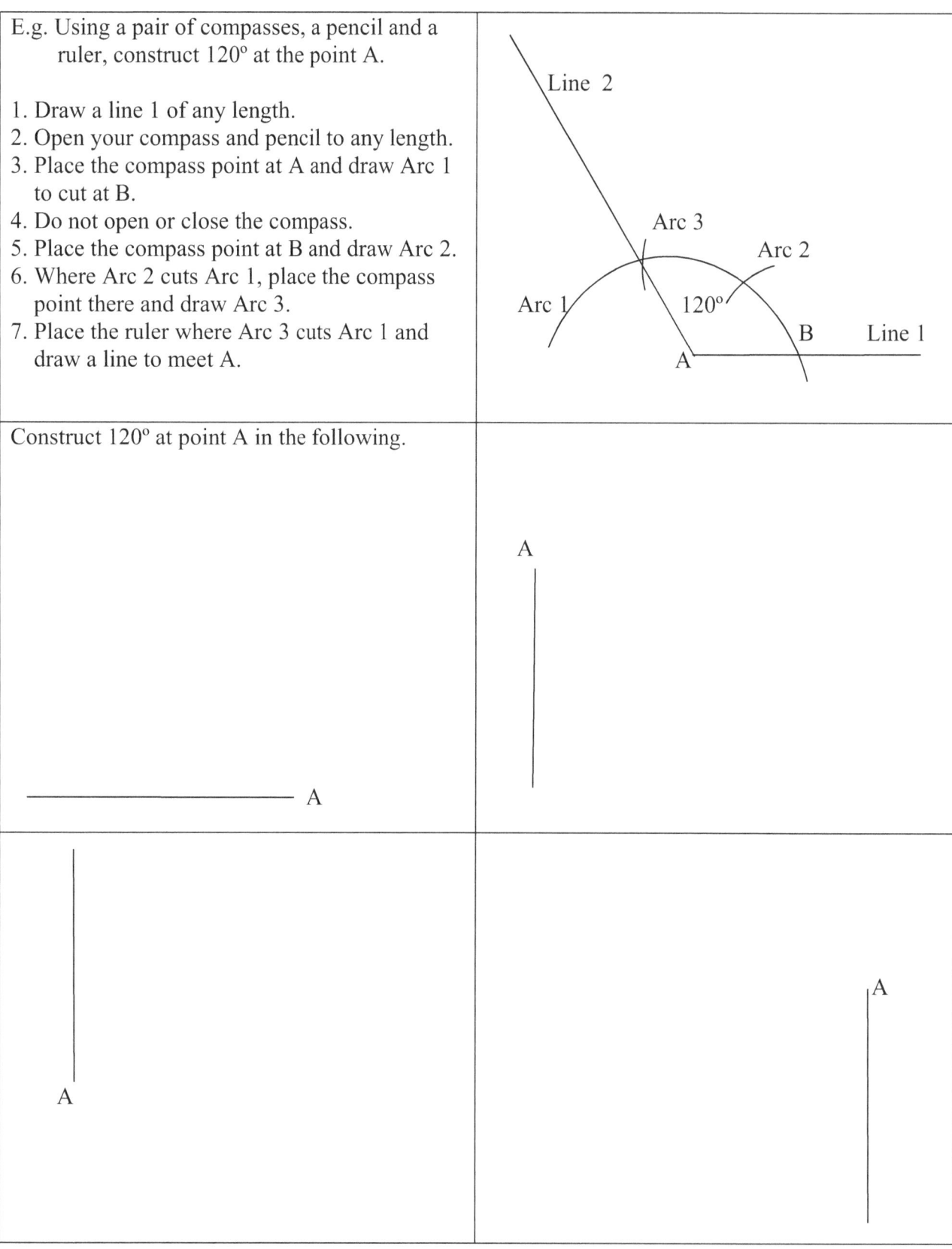

Construct 120° at point A in the following.

A
_____ A

A

A

57

Using a pair of compasses, a pencil and a ruler, construct 120° at point A in the following.

12   Construction of 90º

E.g. Using a pair of compasses, a pencil and a
       ruler, construct 90º at point A.

1. Draw a Line 1 of any length.
2. Open your compass and pencil to any length.
3. Place the compass point at A and draw Arc 1
4. Do not open or close the compass.
5. Place the compass point at B and draw Arc 2.
6. Where Arc 2 cuts Arc 1, place the compass
   point there and draw Arc 3.
7. Where Arc 3 cuts Arc 1, place the compass
   point there and draw Arc 4.
8. Place the compass point where Arc 2 cuts
   Arc 1 and draw Arc 5.
9. Where Arc 4 and Arc 5 cut, place the ruler
   there and draw a line to point A.
10. Use your protractor to measure the angle.

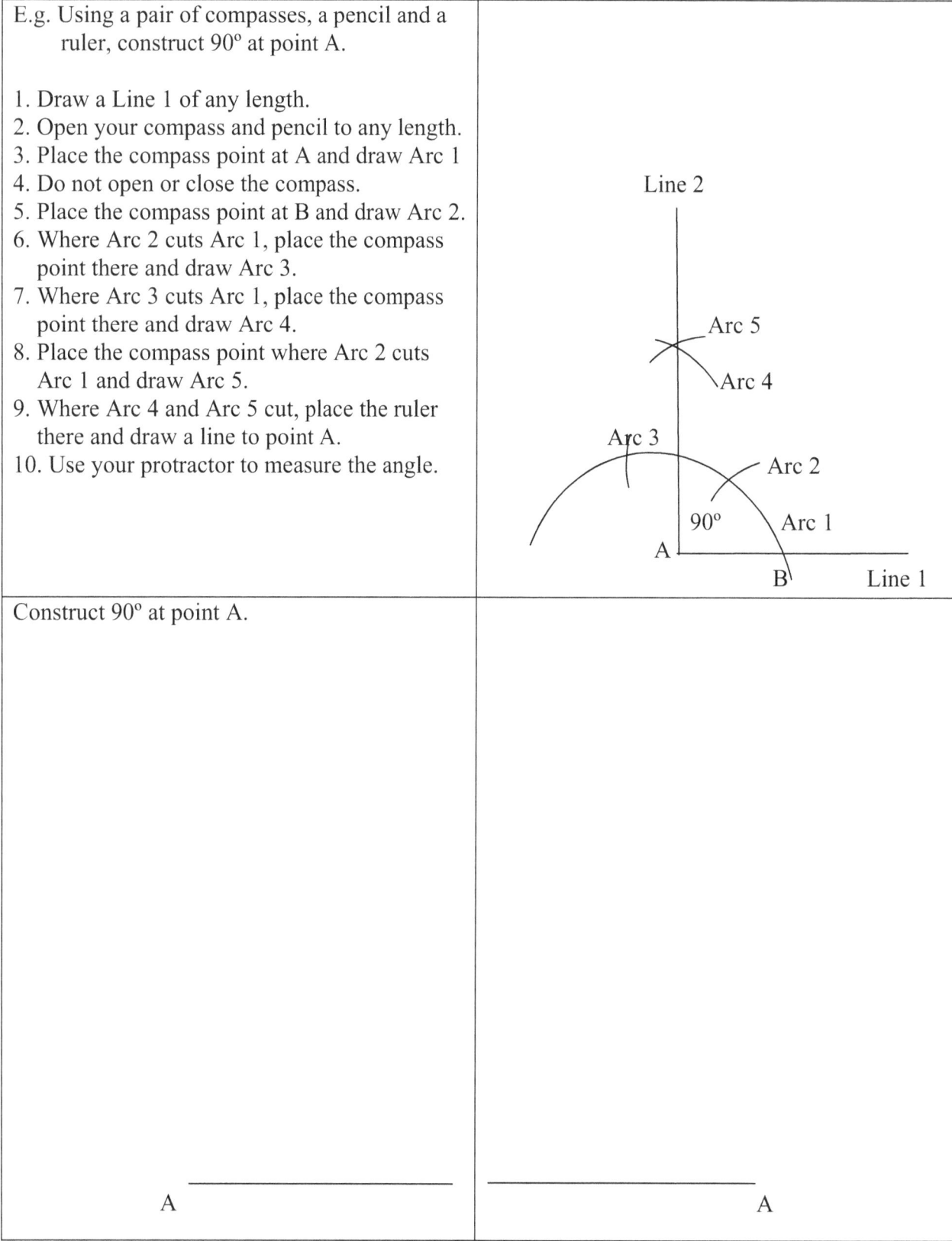

Construct 90º at point A.

_____
A

_____
A

Using a pair of compasses, a pencil and a ruler, construct 90° at point A.

A

A

A

A

Using a pair of compasses, a pencil and a ruler, construct 90º at point A.

13   Construction of 45º

Using a pair of compasses, a pencil and a ruler, construct 45º at point A.

STEP 1: Construct 90º like how you did in the previous chapter. STEP 2: Bisect the 90º like how you bisected angles before.	
A	A
A	A

Using a pair of compasses, a pencil and a ruler, construct 45° at point A.

A

A

A

A

Using a pair of compasses, a pencil and a ruler, construct 45° at point A.

A

A

A

A

14   Construction of two distinct angles on a line

Using a pair of compasses, a pencil and a ruler, construct the following angles.

Construct 90º at A and 60º at B.	Construct 60º at A and 90º at B.
A B	A B
Construct 120º at A and 60º at B.	Construct 60º at A and 120º at B.
A B	A B

Using a pair of compasses, a pencil and a ruler, construct the following angles.

Construct 120° at A and 90° at B.	Construct 90° at A and 120° at B.
A B	A B
Construct 60° at A and 30° at B.	Construct 30° at A and 60° at B.
A B	A B

Using a pair of compasses, a pencil and a ruler, construct the following angles.

Construct 30° at A and 45° at B.	Construct 45° at A and 30° at B.
A B	A B
Construct 30° at A and 45° at B.	Construct 45° at A and 30° at B.
A B	A B

Using a pair of compasses, a pencil and a ruler, construct the following angles.

Construct 60° at A and 90° at B.	Construct 90° at A and 60° at B.
A ———————————— B	A ———————————— B
Construct 60° at A and 120° at B.	Construct 120° at A and 60° at B.
A ———————————— B	A ———————————— B

Using a pair of compasses, a pencil and a ruler, construct the following angles.

Construct 45° at A and 60° at B.	Construct 30° at A and 45° at B.
A _____ B	A _____ B
Construct 60° at A and 60° at B.	Construct 90° at A and 90° at B.
A _____ B	A _____ B

Using a pair of compasses, a pencil and a ruler, construct the following angles.

Construct 30º at L and 90º at M.	Construct 60º at Q and 90º at P.
L M	Q P
Construct 45º at P and 60º at Q.	Construct 90º at G and 60º at E.
P Q	E G

Using a pair of compasses, a pencil and a ruler, construct the following angles.

Construct 60° at A and 120° at D.	Construct 60° at A and 90° at B.
A ———————— D	A ⎮ B
Construct 90° at B and 60° at A.	**Construct 90° at G and 60° at E.**
B ... A	E ... G

15   Construction of a triangle

E.g. Construct a triangle ABC where AC = 6 cm, AB = 5 cm and BC = 5.5 cm.
FIRST: Make a sketch of the triangle and label it. This will be your guide for the actual construction.

SKETCH

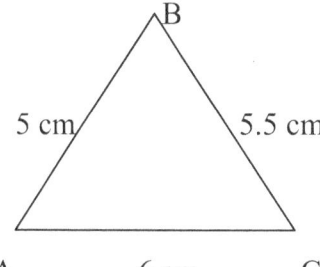

ACTUAL CONSTRUCTION

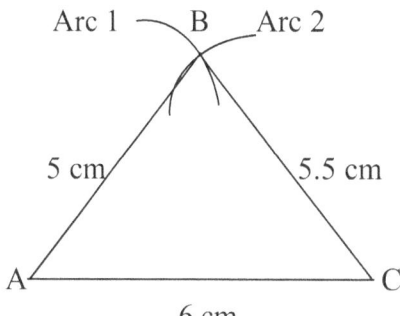

A   6 cm   C
1. Use your ruler and pencil to draw length AC.
2. Open the compass point and pencil point to
   5 cm on a ruler.
3. Put the compass point at A and draw Arc 1.
4. Open the compass point and pencil point
   to 5.5 cm on the ruler.
5. Put the compass point at C and draw Arc 2.
6. Where both Arcs cut, use your ruler and draw
   lines AB and BC.

Construct a triangle LMN where LM = 8 cm, MN = 7 cm and LN = 5 cm. Remember to make a sketch first to show all the information.

Construct triangle PQR where PQ = 7 cm, QR = 7 cm and PR = 7 cm. Remember to make a sketch first to show all the information.

Construct a triangle WXY where WX = 8.5 cm, WY = 5.1 cm and XY = 4.5 cm. Remember to make a sketch first to show all the information.

E.g. Construct the triangle shown in the sketch.

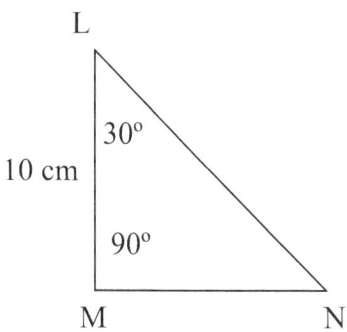

1. Draw a line MN of any length.
2. Construct a 90° at M.
3. Draw the line ML more than 10 cm.
4. Open the compass point and pencil point to a length of 10 cm on your ruler.
5. Place the compass point on M and draw an arc on line ML. This arc means that a length of 10 cm was just cut off from the line ML.
6. Place your compass point on the point where the arc cuts the line ML. Construct a 60° angle at this point.
7. Bisect this 60° to get the angle of 30° at the point L.
8. Draw a line from L through the point that indicates 30° to meet the line MN at N.
9. PRACTICAL: Label the construction below with all the information seen in the sketch.

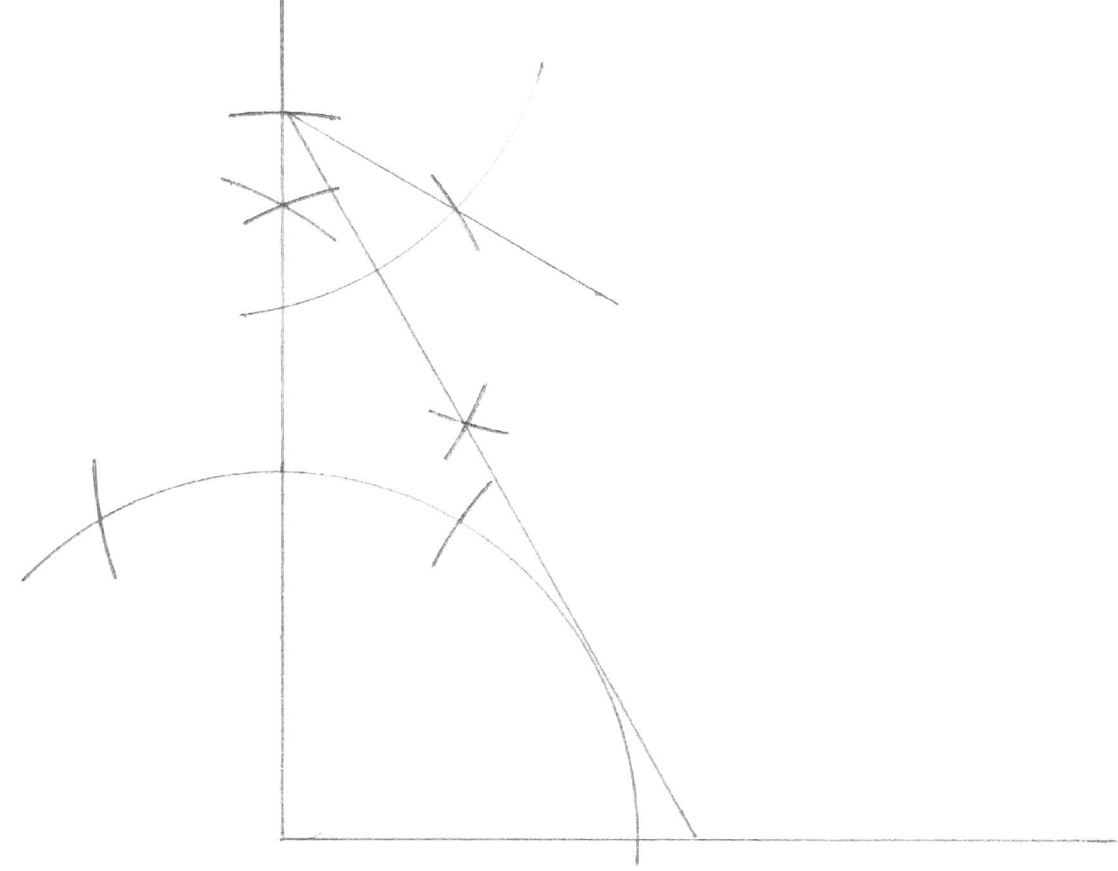

Construct the triangle shown in the sketch.

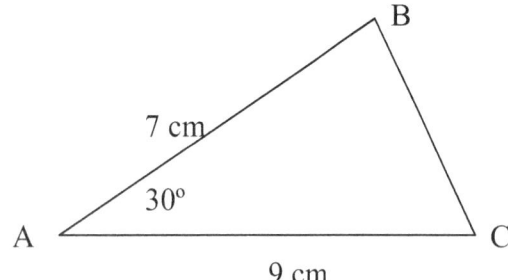

Construct the triangle shown in the sketch.

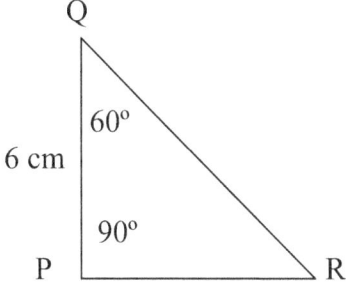

Construct the triangle shown in the sketch.

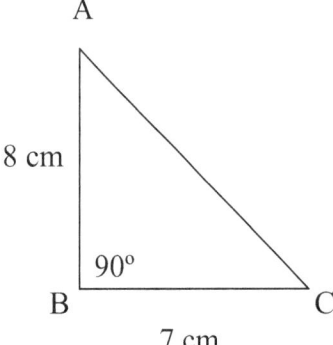

Construct the triangle shown in the sketch.

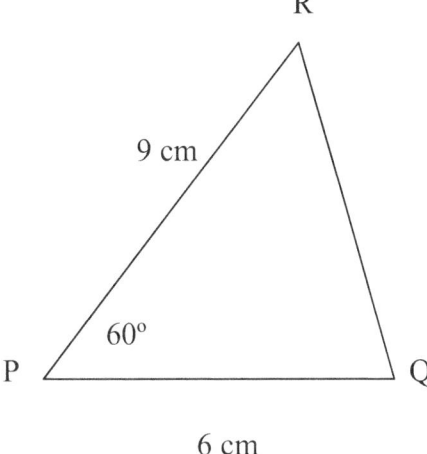

Construct a line from R that is perpendicular to PQ and meets PQ at N.

Construct the triangle shown in the sketch.

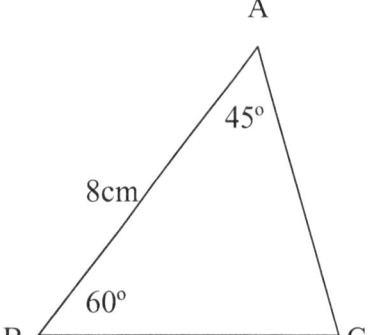

Construct the triangle shown in the sketch.

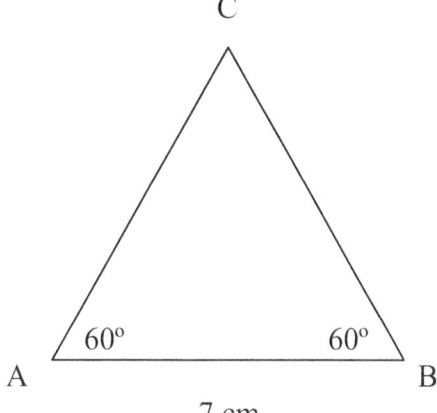

Construct the triangle shown in the sketch.

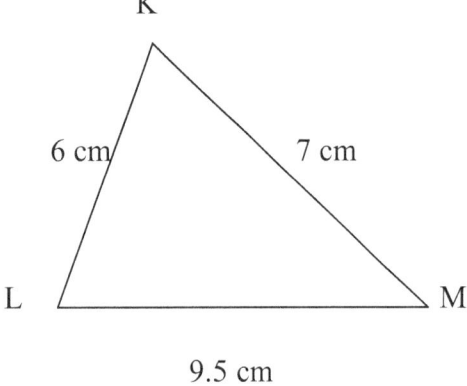

Construct a line from K that is perpendicular to LM.

Construct the triangle shown in the sketch.

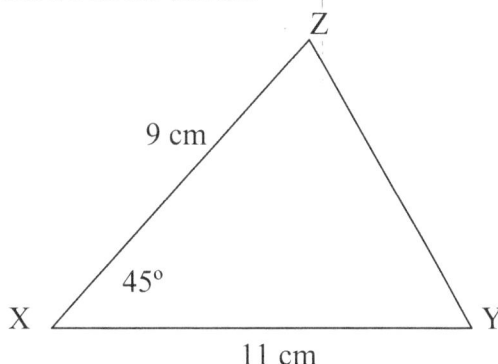

Construct a line from Z that is perpendicular to XY and meets XY at T.

Construct the triangle shown in the sketch.

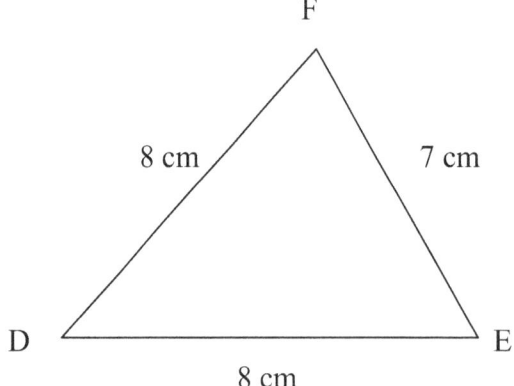

Construct a line from F that is perpendicular to DE and meets DE at T.

16   Construction of a square

E.g. Construct a square WXYX where each side is 8 cm.

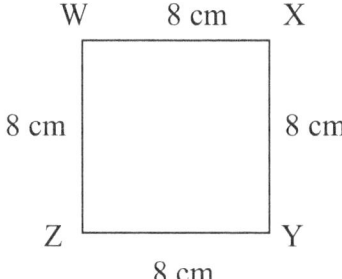

1. Draw the line ZY with a ruler and a pencil.
2. Using a pair of compasses and a pencil, construct a 90º angle at Z.
3. Draw a line from Z to W that is more than 8 cm in length.
4. Open your compass point and pencil point to 8 cm on your ruler.
5. Place the compass point at Z and draw an arc to cut off 8 cm at W.
6. Construct a 90º angle at Y.
7. Draw a line from Y to X that is more than 8 cm in length.
8. Open your compass point and pencil point to 8 cm on your ruler.
9. Place the compass point at Y and draw an arc to cut off 8 cm at X.
10. Draw a line from the arcs at W and X.

FOLLOW THE INSTRUCTIONS ABOVE AND CONSTRUCT WXYZ.

Construct a square ABCD with sides 6 cm.

Construct a square KLMN with sides 5 cm.

Construct a square PQRS with side 4 cm

Construct a square KLMN with side 9 cm.

17   Construction of a rectangle

E.g. Construct a rectangle ABCD with length 10 cm and width 6 cm.

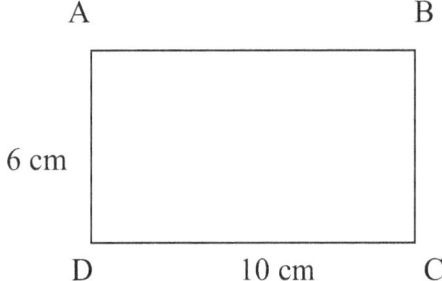

1. Draw a line DC 10 cm in length with a ruler and a pencil.
2. Using a pair of compasses and a pencil, construct 90° at D.
3. Draw a line AD that is more than 6 cm in length.
4. Open the compass point and pencil point to 6 cm on your ruler.
5. Place the compass point at D and draw an arc to cut off 6 cm at A.
6. Using a pair of compasses and a pencil, construct 90° at C.
7. Open the compass point and pencil point to 6 cm on your ruler.
8. Place the compass point at C and draw an arc to cut off 6 cm at B.
9. Draw a line from the arcs at A and B.

FOLLOW THE INSTRUCTIONS ABOVE AND CONSTRUCT WXYZ.

Construct a rectangle ABCD where the length is 8 cm and the width is 6 cm. Measure and state the length of its diagonal.

Construct a rectangle KLMN where the length is 10 cm and the width is 6 cm. Measure and state the length of its diagonal.

Construct a rectangle PQRS where the length is 10 cm and the width is 8 cm. Measure and state the length of its diagonal.

Construct a rectangle ABCD where the length is 7 cm and the width is 6 cm. Measure and state the length of its diagonal.

18   Construction of a parallelogram

E.g. Construct the parallelogram shown in the sketch.

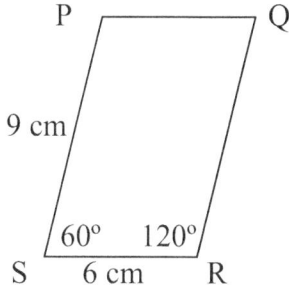

1. Draw line SR 6 cm with a ruler and a pencil.
2. Using a pair of compasses and a pencil, construct 60° at S.
3. Draw a line SP that is more than 9 cm in length.
4. Open your compass point and pencil point to 9 cm on a ruler.
5. Place the compass point at S and draw an arc at P to cut off 9 cm on the line.
6. Using a pair of compasses and a pencil, construct 120° at R.
7. Draw a line from R to Q which is more than 9 cm in length.
8. Open your compass point and pencil to 9 cm on a ruler.
9. Place the compass point at R and draw an arc at Q to cut off 9 cm on the line.
10. Draw a line from the arcs at P and Q.

FOLLOW THE INSTRUCTIONS ABOVE AND CONSTRUCT PQRS.

Construct the parallelogram. Measure and state the length of the diagonal DB.

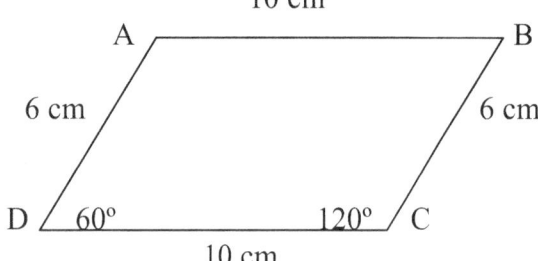

Construct the parallelogram. Measure and state the length of the diagonal ML.

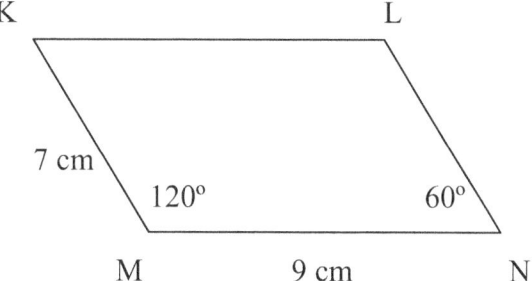

Construct the parallelogram. Extend the line at Y and construct 45° so that angle ZYX is 135°.

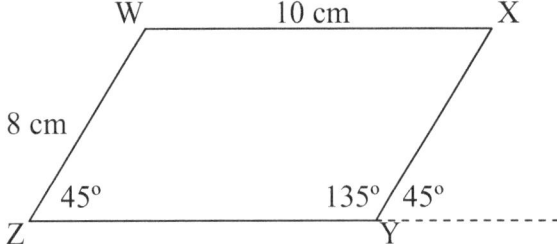

Measure and state the value of the length of WY.

19   Construction of a rhombus

E.g. Construct the rhombus ABCD as shown.

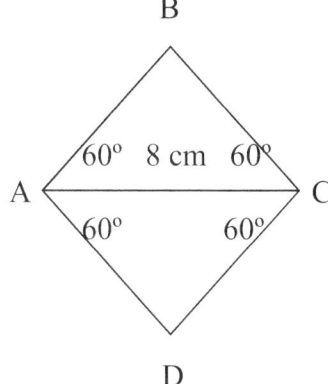

1. Draw the line AC 8 cm in length using a ruler and a pencil.
2. Construct 60° at the top of the line at A using a pair of compasses and a pencil.
3. Construct 60° at the top of the line at C using a pair of compasses and a pencil.
4. Construct 60° at the base of the line at A and at C.
5. Draw the lines AB, CB, AD and CD.

Construct the rhombus in the diagram.

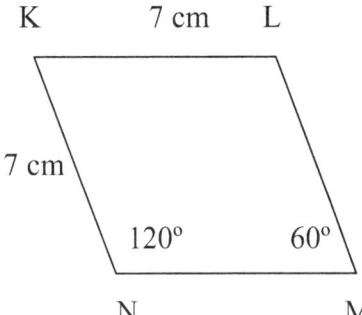

Construct the rhombus.

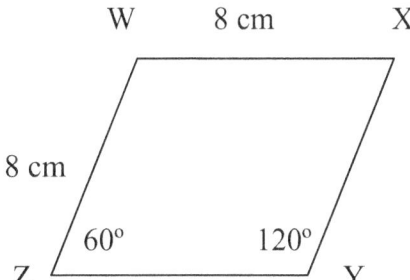

Construct the rhombus.

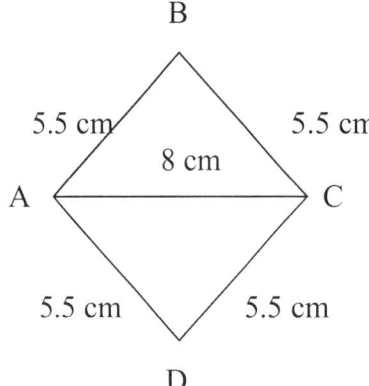

20   Construction of a kite

E.g. Construct the kite ABCD.

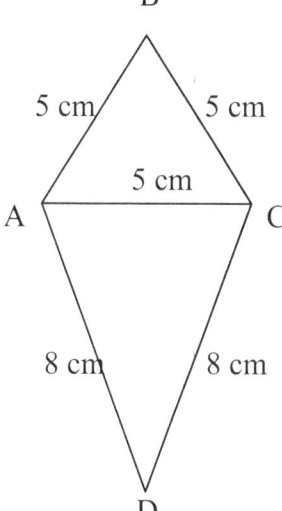

1. Construct the triangle ABC.
2. Open your compass point and pencil point to a length of 8 cm on your ruler.
3. Place the compass point at A and draw an arc at D.
4. Place the compass point at C and draw an arc at D.
5. Use your ruler and pencil to draw the lines AD and CD.

Construct the kite WXYX.

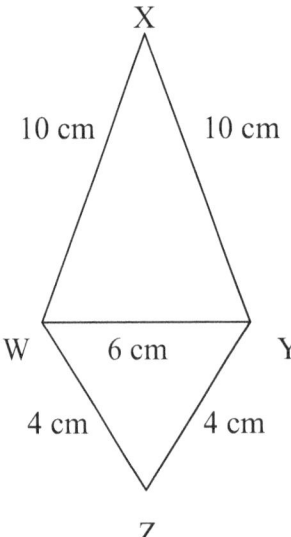

21   Construction of a trapezium

E.g. Construct the trapezium ABCD.

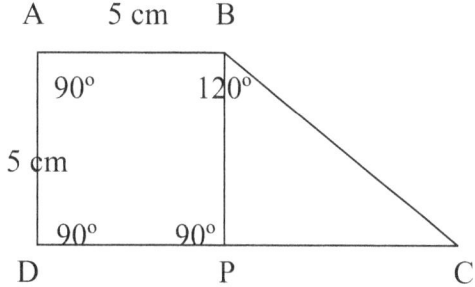

1. Construct the square ABPD.
2. At B, construct 120º using a pair of compasses and a pencil.
3. Extend the line DP and BC to meet at C.
4. State the length of BC.

Construct the trapezium KLMN.

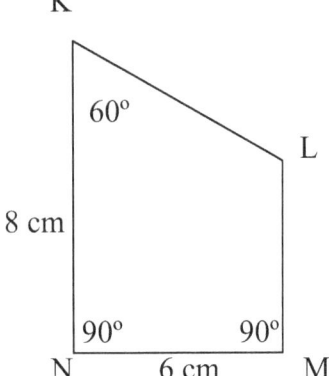

22   Construction of a quadrilateral

E.g. Construct the quadrilateral IJKL.

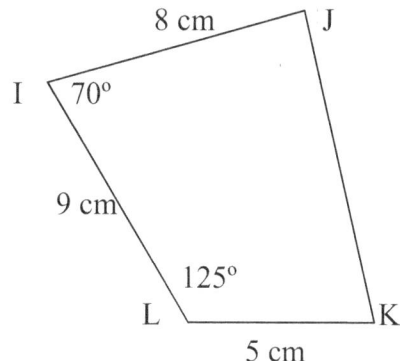

1. Draw the line LK 5 cm with a ruler and a pencil.
2. Use your protractor to draw the angle of 125° at L.
3. Draw the line LI more than 9 cm in length.
4. Open the compass point and pencil point to 9 cm on your ruler.
5. Place the compass point at L and draw an arc at I to cut off 9 cm.
6. Use the protractor and draw an angle of 70° at I.
7. Draw the line IJ more than 8 cm in length.
8. Open the compass point and pencil point to 8 cm on your ruler.
9. Place the compass point at I and draw an arc at J to cut off 8 cm.
10. Use your ruler and draw a line from J to K.

Construct the quadrilateral ABCD.

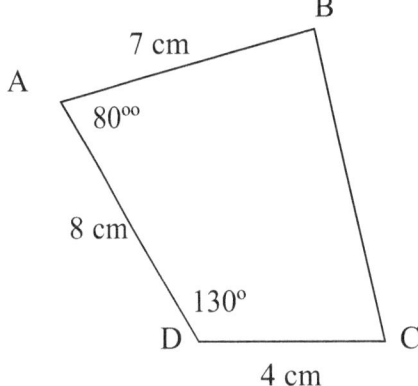

23   Mixed problems

Construct a square WXYZ where each side is 7 cm. Construct 120º at X. Extend the line ZY to form a trapezium WXQZ. Measure and state the length of XQ.

Construct a triangle RST where RS = 8.5 cm, ST = 6.5 cm and angle RST = 90°. On your construction, show the point U where RSTU is a parallelogram.
Measure diagonal RT and state its length.
Measure angle STR and state this value.

Construct a triangle WXY where WX = 9 cm, WY = 7cm and angle XWY = 60º. Construct a line from Y that is perpendicular to WX and meet WX at T. Hence measure and state the length of YT.

Construct a trapezium PQRS where PS = 10 cm, PQ = 8 cm, angle PSR = 90º and angle SPQ = 60º.

Construct triangle LMN, where LN = 9.5 cm, LM = 7.5 cm and angle MLN = 45°. Construct a line from M that is perpendicular to LN.